MGRS EXPLAINED

MGRS EXPLAINED

The Military Grid Reference System in Simple Terms

AARON OVERTON

Heatherstone Knowledge Works

Dedicated to
SFC Timothy Bradley, TNSG

When in doubt, ask yourself, "What would Bradley do?"

CONTENTS

PREFACE

The MGRS coordinate system is a vital part of land navigation and location coordination during operations. However, information online about the system is not particularly easy to understand.

After reading many sources, it was still confusing in some areas. It took looking carefully at actual maps, and especially the online application at https://map.army, then doing more research to explain findings, to get to a solid understanding. In an effort to produce better training on the topic, and also to make sure I really understood it myself, I wrote this primer. I set out to also include a lot of examples with images to help illustrate the points.

There is some geometry and trigonometry in here, when explaining distance and azimuth calculation from coordinates. Hopefully, there is enough explanation that even those without much or recent math background to still understand.

Due to my active involvement with the Tennessee State Guard, the locations within are all relating to Tennessee. Of course, all the guidance applies anywhere on the planet. OK, maybe not at the poles, but most of us don't want to go there anyway.

I enjoyed writing it. I hope you enjoy reading it and learn a lot.

CHAPTER 1

Introduction

The Military Grid Reference System (MGRS) is the standard use by all NATO military forces to locate any point on Earth down to one meter locations. This is called a "geocode" - a unique identifier for any location or object that is short and can be fairly easily read by humans.

An MGRS coordinate consists of three parts:

- A grid zone designator (GZD), consisting of a number and a letter
- A 100km square identifier, consisting of two letters
- A pair of two numbers, designating an easting and a northing, always presented with equal precision

For example, consider an MGRS coordinate, per a screenshot from map.army on the next page. The front steps of the "Triple Nickel" - that is, building 555 on the Tennessee Army National Guard base in Smyrna, is this: **16S EE 44496 85049**

16S is the GZD. Any location in Tennessee is in 15S, 16S, or 17S. Roughly the eastern third of Tennessee is 17S. Only a very small part, mostly the western part of Memphis, is in 15S.

EE is the 100km grid square. There is a pattern to how these squares are labeled, which we will cover later.

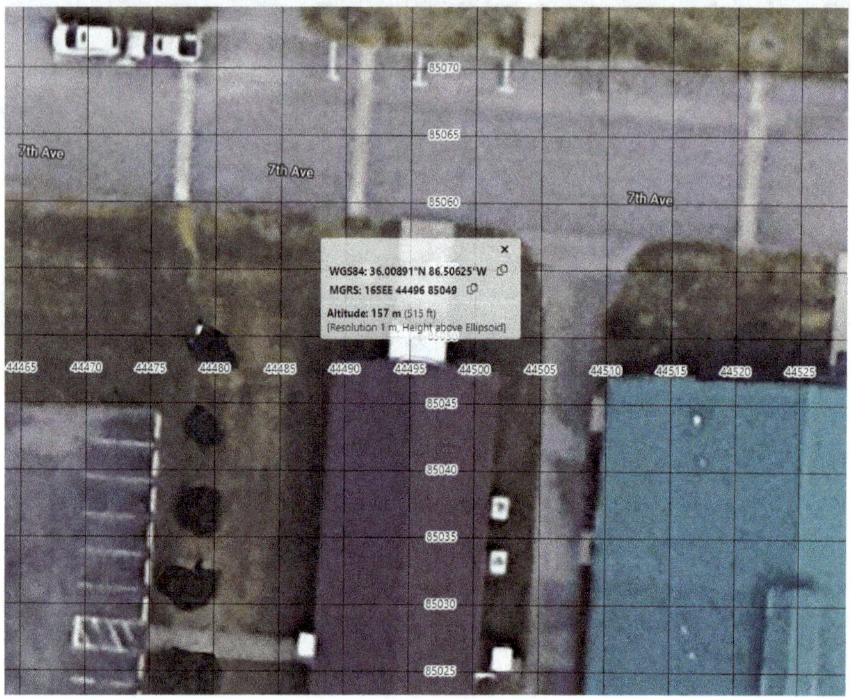

44496 is the easting, which is how many meters east of the west edge of square EE it is to this location. Similarly, 85049 is the northing, which is how many meters north of the south edge of square EE it is to this location. Because each number is five digits, these numbers represent precision down to 1 meter.

There is more to understand in each of the parts. We will start with the details of grid zones.

CHAPTER 2

Grid Zones

The lowest precision portion of an MGRS coordinate is the grid zone designator (GZD). It is made up of a number, from 1 to 60, followed by a letter, A to Z, but skipping I and O because those two letters look too much like one and zero. Fortunately, English has 26 letters and we only needed 24 of them.

WGS84

In grade school, most of us were taught at some point about another geocode, latitude and longitude. As a quick refresher, longitude lines go from the physical north pole to the physical south pole. There are 360 degrees of longitude, though we refer to 0° as the prime meridian on the north-south line that passes through London, England. We then use 180 degrees each to the west and east of that line.

Latitude lines are drawn as circles running east and west around the globe. There are 180 degrees of latitude, but similar to longitude, we refer to 0° as the equator and use 90 degrees each to the south and north.

The current standard for latitude and longitude is called the World Geodetic System and was established in 1984, so is abbreviated WGS84. This is the system used by the Global Positioning System (GPS). Looking at Triple Nickel, we see that the front steps of the building are located at 36.00891°N 86.50625°W.

If you have seen GPS locations in degrees, minutes, and seconds, you'll notice this location doesn't do that. Rather it uses decimal fractions of a degree. This is nice because while we think of time with minutes and seconds, distances aren't quite so easily understood broken into groups of 60.

GZD DESIGNATIONS

The reason to remember latitude and longitude with MGRS is that grid zones are drawn using them. Let's start with the first part of the GZD, the number.

Starting at 180° longitude and going east, every 6° of longitude gets a number, starting with 1. Since the full set of longitude is 360 degrees, we get 60 vertical bands or columns, numbered 1 to 60. Tennessee is in mostly in columns 16 and 17, except for the most western part of Memphis that is just inside column 15.

For the horizontal bands or rows, we use the letters A through Z, except I and O. We will ignore for the moment the areas around the poles except to note that the 10° south of 80°S are the south polar region and use letters A and B and the 6° north of 84°N are the north polar region and use the letters Y and Z.

Starting at 80°S and looking northward, every 8° of latitude is given a letter, in order, starting with C (again, skipping I and O). If you've been adding up the degrees, you might note that with 10° in polar south, 6° in polar north, and 8° per latitude band, the twenty letters we have left from C to X totals 10 + 6 + (8 * 20) = 176°. That means there are 4 extra degrees somewhere - those are included in band X, so instead of 8°, the X band covers 12°.

All of Tennessee is in the S row, which we can see from this map with grid lines showing the GZD and 100km square coordinates in the overlay.

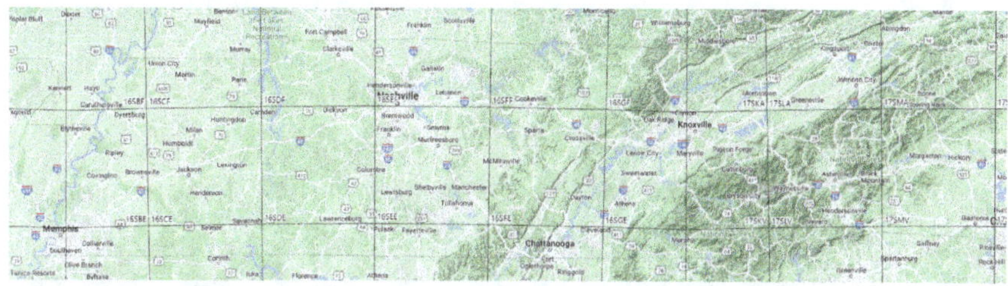

Next, we will look at the system for designating the 100km squares (what we will frequently call "grid squares").

CHAPTER 3

Grid Squares

The second part of an MGRS coordinate is the grid square designator. In the MGRS coordinate for the front steps of Triple Nickel in Smyrna, 16S EE 44496 85049, the letters EE are the grid square. The same letter pairs can be found repeatedly on the globe, but have been designed so that they will not be found sufficiently near each other to cause confusion. Not to worry, there are other ways to get confused about these.

One such confusion is the difference between "MGRS-New" and "MGRS-Old". New is a relative term, however, as the "new" version, also called the "AA" scheme, has been in use since at least 1984 and probably much earlier. Online references as to when this transition happened are elusive.

The first letter is a column and the second letter a row. The columns are slightly more challenging to understand because of how longitude lines are closer together the farther north or south one is away from the equator.

LONGITUDE, AGAIN

At the equator, longitude lines are at their farthest apart and have the same separation as latitude lines. This is about 111km (or 69 miles) per degree of longitude. As we look farther north or south, longitude lines get closer as they all converge on the poles. At 36°N, the latitude for Triple Nickel, the distance between longitude lines is about 90km (or 55 miles).

This means that in each Grid Zone north of the equator, the north end of the grid zone is narrower than the south end. Row S, where all of Tennessee is located, is at 30°N at the south and 38°N at the north. Each degree of longitude is therefore about 96km at the south edge and about 87km at the north edge. Since latitudes are always equally distant, each degree is about 111km.

COLUMN LETTERS

The letters used in for columns are different depending on the number in the GZD. Every third column, starting with column 1, so 1, 4, 7, 10, etc, use A through H. Every third column, starting with column 2, so 2, 5, 8, 11, etc, use J through R (except for O). Every third column starting with column 3, so 3, 6, 9, 12, etc, use S through Z.

However, because of the shrinking distance between longitude lines, as we go farther north or south, we don't need all the letters to fully cross a grid zone. We can do the math on this: for column 16, containing the Triple Nickel building, we use letters A through H. At the south edge (at 30°N latitude), we know that 6° of longitude is about 96km, so that south edge is 576km long. Since each grid square is designed to be 100km square, we only need six letters. A-H is 8 letters, so we need to know which ones we will use.

The squares are arranged so that the middle point of each grid zone is the divider between the four letters west and east of that line. So normally, if there were exactly 8 100km squares, we would put A, B, C, and D to the west. But we only have half of 576km on the west of the middle line, or 288km, so we don't need A. Similarly, on the east, we don't need H.

The north edge of column 16 is narrower, at 522km. That is still wide enough that we use letters B through G at the north as well. You can see this on this map around Jackson, Mississippi, where Grid Zone 15 and 16 meet:

ROW LETTERS

Row letters are a bit easier to understand because latitude lines are equidistant, or don't converge. Starting at the equator and going north, every 100km gets a letter. As with the letters used for columns, the starting letter depends on the Grid Zone column. We start with A for odd-numbered zones and F for even-numbered zones.

These letters continue north until the letter V (except I and O), then start again at A. South of the equator, the letters count backward as we go south, so the first row south of the equator in odd-numbered zones is V and in even-numbered zones is E.

DON'T WORRY

This is a good place to point out that a 100km square is a really big area. That is 10,000 square kilometers, which is about 2.5 million acres in handy homeowner terms. Here is a map showing the 16S EE grid square in which we find Triple Nickel. The large square is the entire grid square. The small square is a rough approximation of the entire Tennessee Army National Guard base in Smyrna.

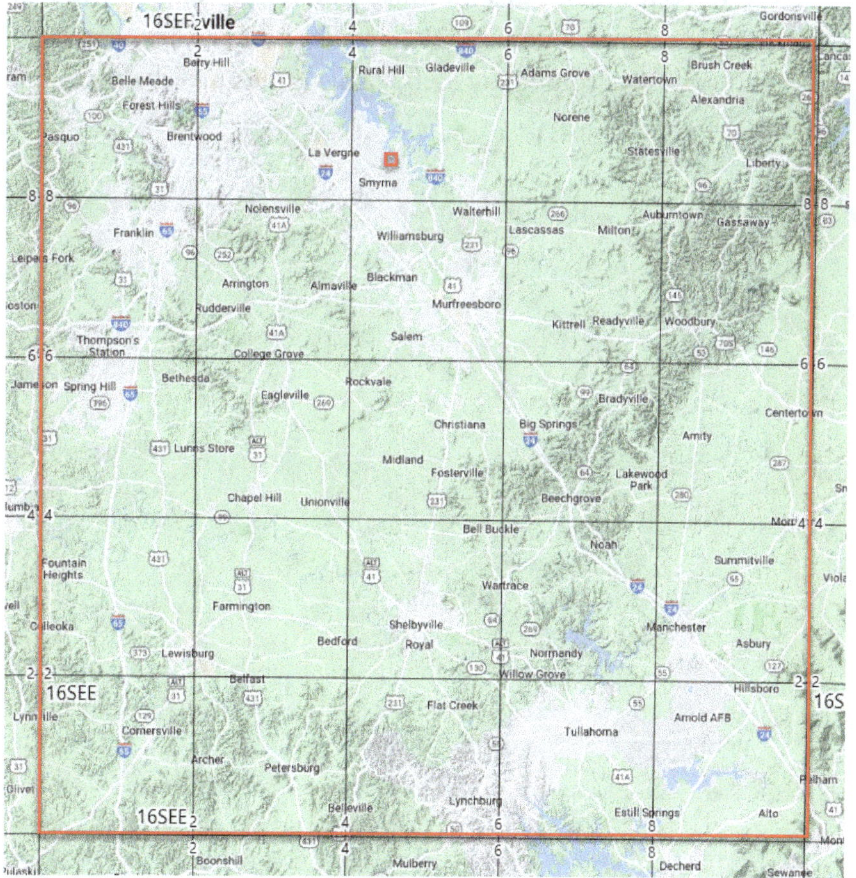

Anyone who doesn't fully understand everything that's been described so far, until we are conducting operations that span the entire state of Tennessee, grid squares aren't going to matter much. We can do an exercise at the Smyrna facilities and every coordinate will be in 16S EE, so we will always just use the numeric third part of the coordinates.

Also, maps with MGRS coordinate grids will include the Grid Zone and grid squares. A typical USGS topographical map at 1:24000 scale will have 1km markings. It is more common that such maps are entirely within one grid square and rare that one lands on an intersection of four grid squares.

If you've been finding this information a lot, it's okay, it's mostly background most people won't use often. The next part is when it gets both easier and more commonly useful. That is the numerical location.

CHAPTER 4

Numerical Locations

The third part of an MGRS coordinate is the numerical location. Once again, let's look at the coordinates for the front steps of the Triple Nickel building at the Tennessee Army National Guard base in Smyrna: **16S EE 44496 85049**

MEANING

The two numbers are called an "easting" and a "northing", and are listed in that order. The easting is a measure of how many meters east the location is from the west edge of the grid square. For Triple Nickel, those steps are 44496 meters, or about 44.5Km east of that edge.

Similarly, the northing is a measure of how many meters north the location is from the south edge of the grid square. For Triple Nickel, it's just over 85km. Here's a map of grid square 16S EE showing (roughly) how those measurements work:

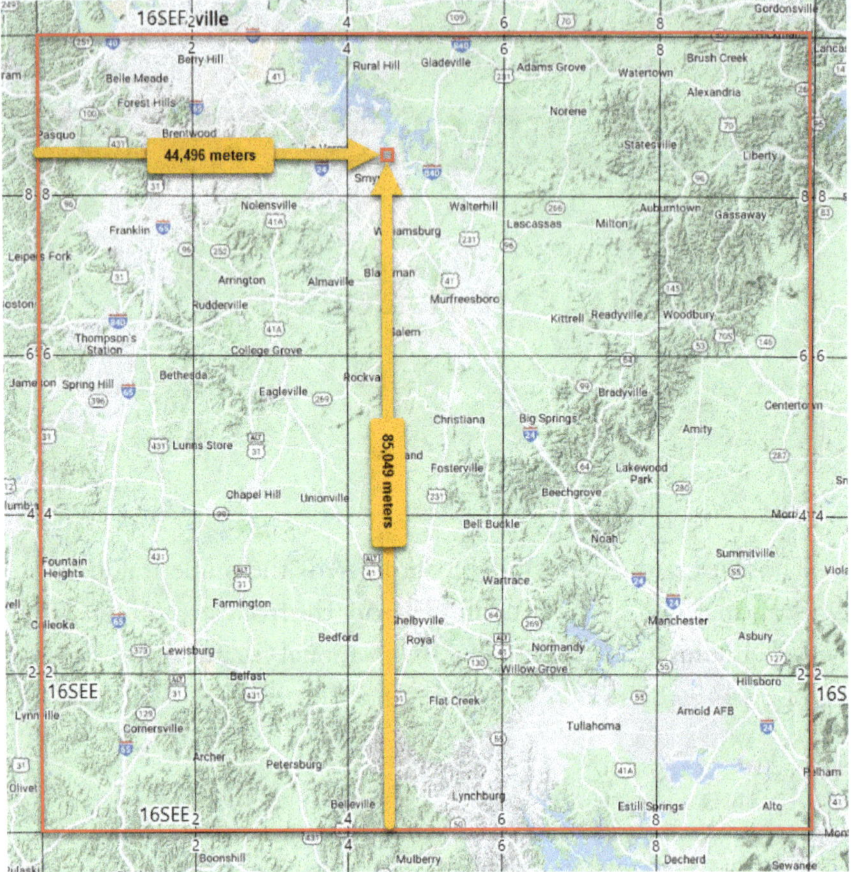

This is so much simpler than understanding Grid Zones and grid squares. That is where the use of this system really shines when being used in the field.

PRECISION

The numerical location may or may not include a space between the easting and the northing numbers. We've been looking at coordinates with spaces between the parts to make it easier to read and understand, but in practice, all the spaces will often be left out. This matters most with the numerical location because we usually don't need or even have full precision down to one meter.

The easting and northing are always used with the same number of digits. In our examples so far, we are using five digits for each and recognizing that these are representing meters. If we have a 1:24000 scale map, a one meter difference in the real world is .04mm on paper. To get a sense of how small that is, here are the smallest of the tips on a set of "Micro-Line" pens:

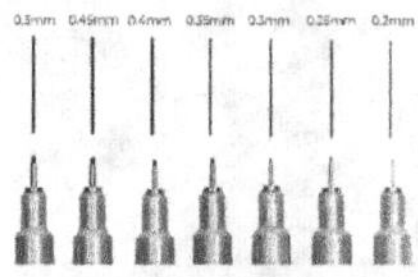

The smallest draws a line that is 0.2mm in width, which is five times the distance of how a one meter, real-world distance is shown on a 1:24000 scale map.

When we don't have that much precision, we use fewer numbers for the easting and northing, but again, the same number of digits for each. When we use only four digits, Triple Nickel's numerical location becomes 4449 8504, and it represents a 10m square location instead of a 1m square location. Those coordinates are here on a satellite map:

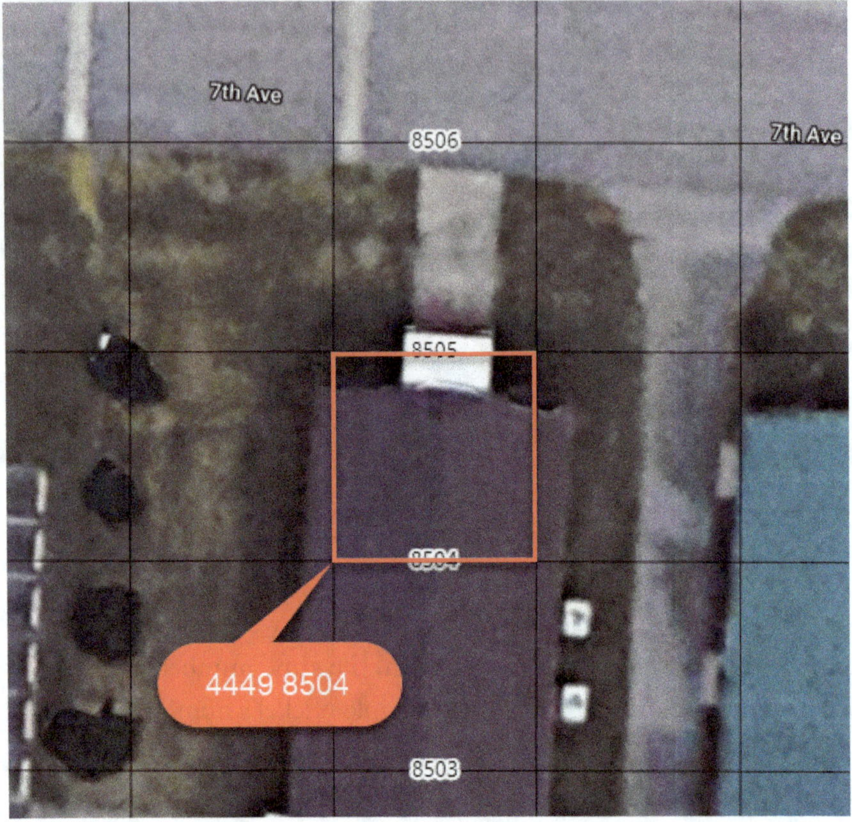

As we can see, if we were only given four digits each for easting and northing, we could still easily find our way if the goal was, "meet on the front steps of Triple Nickel". If we had only three digits each, it would be 444 850. Now we would be looking at a 100m square, which would make it a little harder to find, but not impossible. There would be a little walking around to find the right building, but there are only two buildings with entrances in the 100m square:

PADDING

When we fill in the digits of a small number so that it has the same number of digits as a larger number, we call that "padding". Padding is also used with letters or spaces in other systems and comes up regularly in software development, for example. Here, we use padding to make sure each of the two parts of the numerical location have the same number of digits.

Take this example: There is a building with a blue roof in a remote part of Panama at coordinates 17PNK 00647 12312. Look at just the easting, 00647. It is close to the west edge of the 17PNK grid square, 647 meters. If we only used "647" and jammed all the parts of the coordinate together, we'd get 17PNK64712312. With only eight digits, the easting and northing could get interpreted as "6471" and "2312" - a stand of trees in the middle of a field roughly 11.6km away.

TRUNCATION

These examples also show us another important point, which is "truncate, not round". If we were rounding the numbers to only three digits, the easting would go from 44496 to 445. With truncating, we just drop the last numbers, so 44496 becomes 444.

Since the coordinates always tell you the south-west corner of the area containing the location, rounding would have us looking in the wrong 100m square. Another way to put this is that by truncating, the higher precision numerical location is contained within the space referenced by the lower precision numerical location.

With the numerical locations, we now understand all three parts of an MGRS coordinate. This is enough to find locations in the real world or on maps with MGRS coordinate grids. With some work, we could figure out latitude and longitude as well, though it can be a bit tricky because of what we discovered about how grid squares work with longitude lines.

GRID ZONE BOUNDARIES

Everything so far has been based on the assumption that grid squares are 100km on each side, which means full precision numerical locations would range from 00000-99999. Unfortunately, that pesky longitude thing where longitude lines get closer together comes into play when looking at eastings near grid square boundaries.

Let's look at another example:

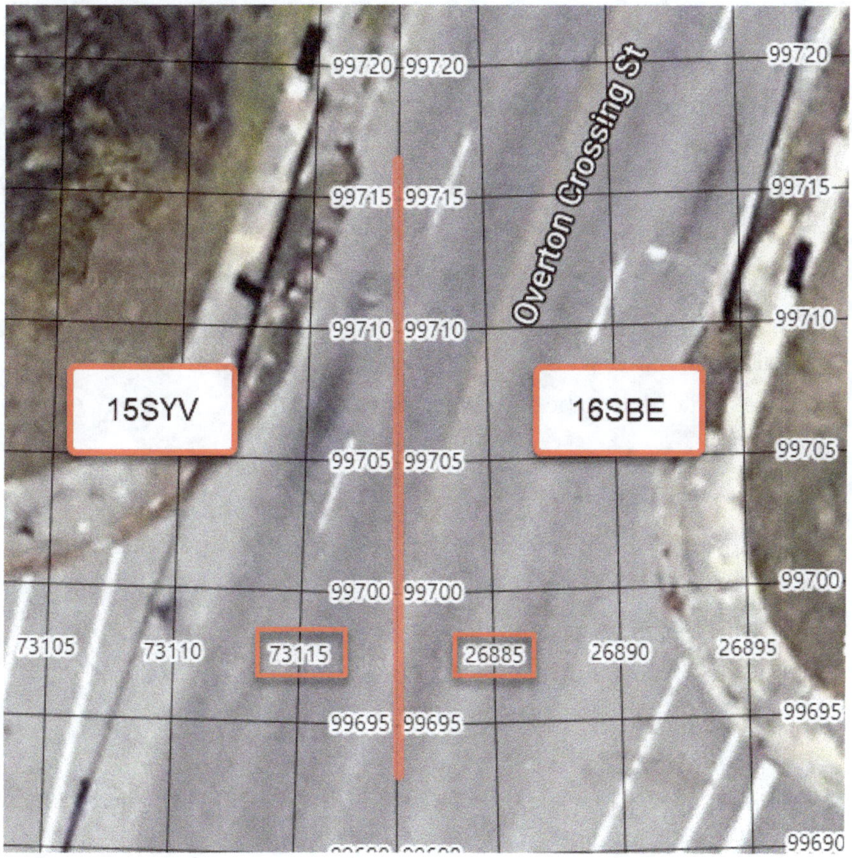

This is along the boundary between grid square 15SYV, the most eastern part of column 15, and grid square 16SBE, the most western part of column 16. The fact that this happens to land on Overton Crossing St was purely accidental, but a happy coincidence.

The grid is showing lines every five meters. The last five meter increment in 15SYV is 73115 and the first five meter increment in 16SBE is 26885. The numerical location eastings in between don't go up to 99999 or down to 00000.

Before, we discussed how the column letters are assigned. Within each column of a grid zone, the grid squares are lined up with the center of the column and apportioned equally on east and west of that line. When we reach the east and west edges of the column, we stop. This means that the most eastern and most western grid squares are not full squares, but are cut down more and more the farther north or south we go.

The main thing to remember is that when operating near grid zone boundaries, some of the other rules might not apply. We looked at how large a grid square is and grid zones are even larger, so fortunately, in most circumstances, with Overton Crossing St being one of the exceptions, we don't need to be concerned with this.

CHAPTER 5

Navigation

The primary use for MGRS coordinates are to quickly find locations on a map or to transmit location information to others quickly, easily, and precisely. We can also use them, with a little bit of fancy math, to navigate between two points.

What we will discuss first applies for two points in the same grid square. If two points are not in the same grid square or even in the same grid zone, it gets much more complicated, so we are going to just ignore that for this discussion. If you master everything else in this book, working out how to apply it when crossing grid squares or grid zones will be readily achieved on your own.

OUR EXAMPLE

For a practical example, we will look at two locations at the Tennessee Army National Guard facility in Smyrna. We will assume we are at the RAC Crew Shack at coordinates 16SEE 44022 85676 and we need to get to the largest building in a group of green-roofed buildings in a training area at coordinates 16SEE 44412 86231. We're operating in the same grid square, so for the rest of this example, we will leave off 16SEE and just think about the eastings and northings.

We need two things: an azimuth and a distance. An azimuth is the angle in degrees we will follow on a compass. With those, we can walk outside, look at our compass, and walk in that direction for the distance we figure out, and we will be at our destination. Except for all those trees and stuff in the way - but there are ways to work around that.

DISTANCE

We will start with distance because that is easier. We will use a basic geometry formula, the Pythagorean theorem, to get the distance. If you are "not a math person", don't be concerned (and also stop saying that, because the best way to make sure you are not a math person is to keep saying it.)

The Pythagorean Theorem applies to right triangles. That is, triangles that have a right angle (a 90° angle.) The basic formula is:

$$a^2 + b^2 = c^2$$

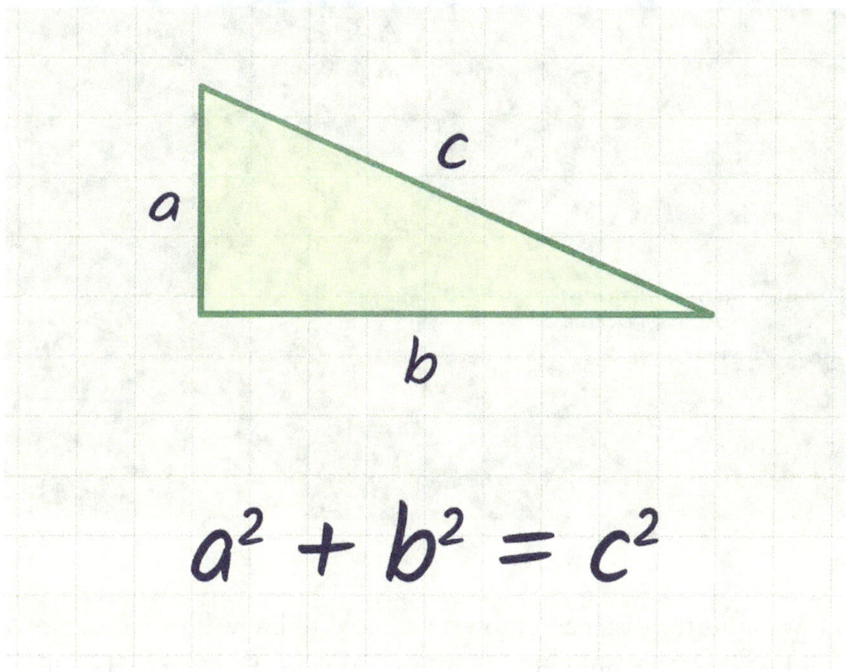

$$a^2 + b^2 = c^2$$

Let's look at our map again, but with a triangle drawn on it:

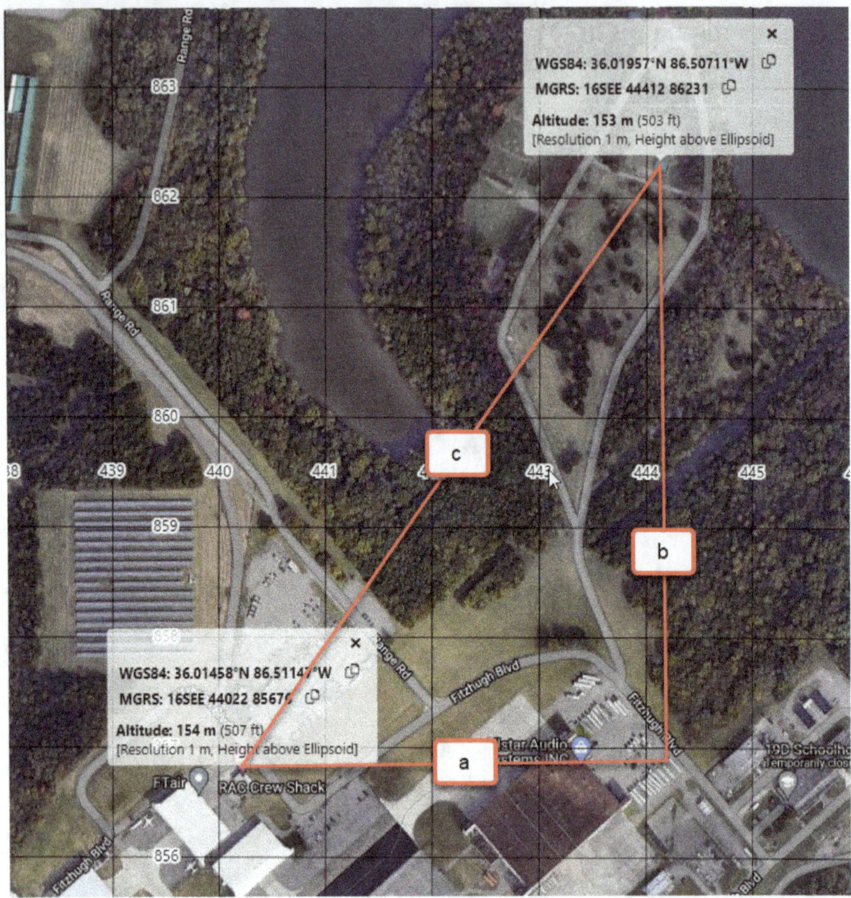

We want the distance, c, so we just need to know how far each of a and b are. MGRS numerical locations come to the rescue. The difference in eastings between our start and finish is a. That is, 44412 - 44022 = 390m. Similarly, the difference between the northings is b. That is, 86231 - 85676 = 555m.

So now we know that $390^2 + 555^2 = c^2$. Basic algebra says we can figure out c by taking the square root on both sides of our equals sign, so now we are just punching a couple numbers into a calculator. Once we do, there it is, 678m is our distance between these two points. Next we can look at the azimuth.

AZIMUTH

Getting the azimuth is a little harder because we have to tap into trigonometry. Fortunately, once you have the basic idea, we can just go back to our calculator and throw some numbers in, do one small adjustment, and we have the answer. Let's start with explaining the math bit, where our helpful formula is the "inverse tangent" or "arctan".

MATH FOR FUN AND PROFIT

Here's a diagram that shows labels for how we think about the sides of a right triangle in relationship to an angle we care about.

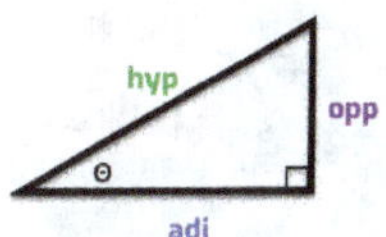

The abbreviations are for "adjacent", "opposite", and "hypotenuse" (the fancy term for the longest side in a right triangle.) In this diagram, the Θ symbol is the Greek letter "theta", and is the angle. The formula we want is:

Θ = arctan (opp / adj)

Arctan will tell us the angle we need using information we already have from calculating distance, the adjacent and opposite sides. Let's look at this again on our map, with new labels:

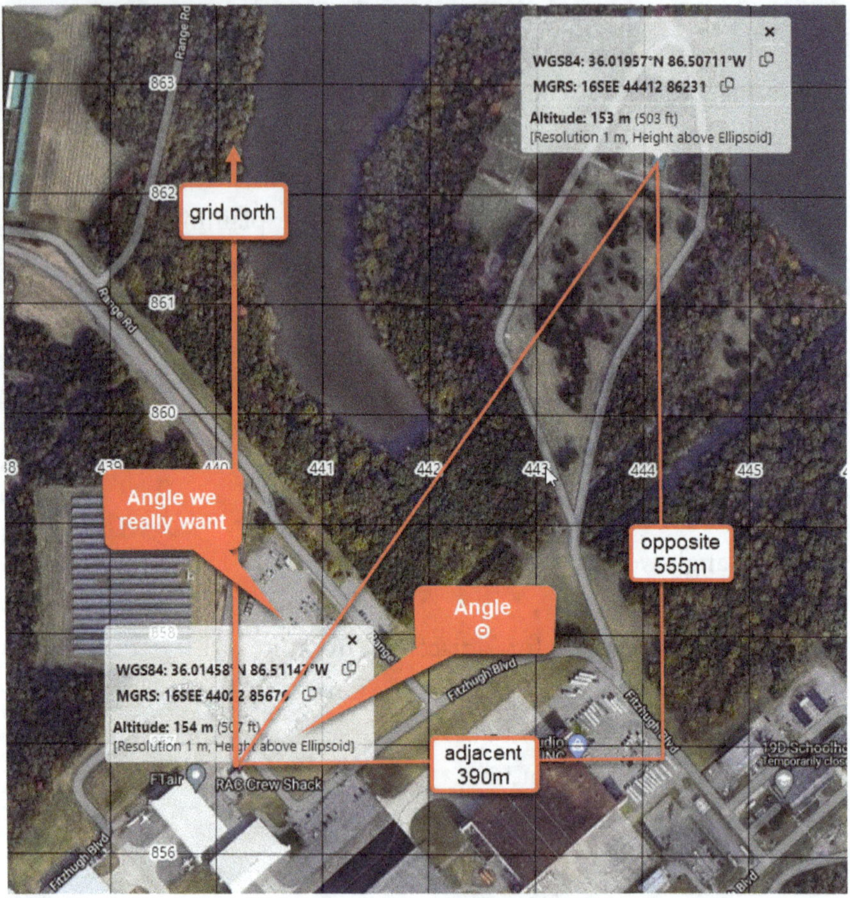

We want the angle labeled Θ, in degrees, and we have our formula using arctan. On calculators, arctan might show as $\tan^{-1}$. Use an iPhone's calculator app, in landscape view, to get all the extra buttons.

Type in "555", then divide, then "390". Press 2nd (if it's not already highlighted), then $\tan^{-1}$. The result is approximately 55°. We aren't quite done yet, but we're close. Two things to note from the image above: First, the button in the bottom left, "Rad", means "radians". If it shows "Deg", then you will also see "RAD" in the upper left corner and the result you'll get is in radians instead of degrees. This isn't helpful for us here. So make sure that button shows "Rad".

Second, we had to press the "2nd" button to get a different set of calculations. The arctan ("tan^{-1}") button is the secondary function of the "tan" button, so we pressed "2nd", then arctan.

Now that we know 55° is that angle, we should also note that for our compass direction or azimuth, we need the other angle shown in the map above. One more simple piece of knowledge to apply here is that the two parts of a right angle add up to 90°, so 90 - 55 = 35°, which is our actual (grid north) azimuth.

It is going to matter if your general direction is generally NW, NE, SW, or SE to determine how you take the angle and adjust it to get your azimuth. Try drawing a few directions to see which angle you are getting and how you need to adjust it to get your azimuth, as an exercise to see if you are getting all this.

If we walk outside of the RAC Crew Shack and go 678m at 35° on our compass, we're there, right? Well, almost. Unfortunately, we need to consider the three norths.

CHAPTER 6

Using with a Map

THE THREE NORTHS

Above, we said "our actual (grid north) azimuth". In land navigation, we have to consider three norths: grid, magnetic, and true.

True north points to the geographic northern-most point of Earth. In theory, this is a fixed location, although it does move a small amount over time as the planet wobbles a bit. Since that wobble takes 25,000 years to make a loop, we can treat true north as a fixed point.

Grid north, as it has been designed, changes not at all and is almost the same as true north. The difference is usually so small that the accuracy of a handheld compass wouldn't be able to show it.

Magnetic north, which is what our compasses target, varies based on our location and isn't even consistent. Also, the magnetic north pole moves over time as the Earth's magnetic sphere shifts. In the last couple of decades, the magnetic north pole has moved an average of 34 miles every year. GPS satellites are being updated every six months to account for the change.

In fact, our compasses don't even point to the magnetic north pole, but only show us the direction of flow of the magnetic field at our current location. Since the magnetic field shifts and doesn't even follow the nice, straight longitude lines on our maps, it isn't even predictable. Still, this is what we have when walking around in a situation where we don't have our gadgets that use satellites, electricity, and all the rest.

DECLINATION

To help with this, maps are printed with true north and magnetic north, and when there is an MGRS grid overlay, with grid north as well. The date of the map is also printed, which helps because the older the map, the greater the chance that the magnetic north shown is incorrect. The difference between magnetic north and true (or grid) north is called the "declination".

On a map, the declination is usually shown like this:

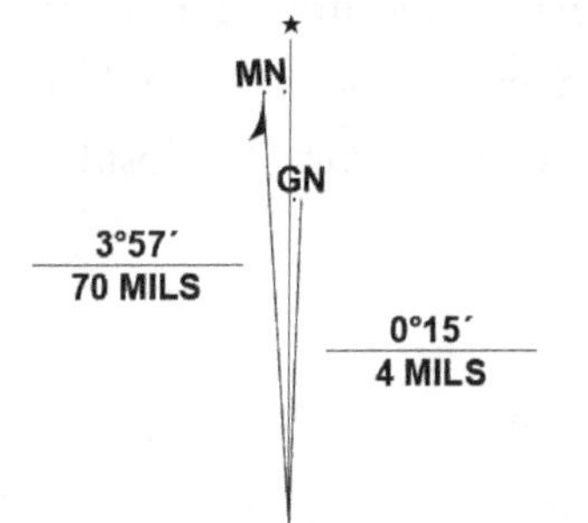

This is the declination for the location in our example. The star represents true north, GN represents grid north, and MN represents magnetic north.

The declination is shown as about 4° W. The 57′ is minutes, and there are 60 minutes in a degree, so it's okay to round to 4°. The grid north declination is only 15′ E, or a quarter of a degree. Or total magnetic declination is about 3.75° W.

When the declination is west, we want to add that amount from our grid azimuth to get our magnetic azimuth. The way to remember that is

"West is best, east is least," so west declination makes the number bigger and east declination makes the number smaller.

Finally! We know our actual azimuth to use on the compass. 35° grid + 4° declination = 39° on our compass. (Why did we round it? Try your compass - are you able to tell the difference of a quarter degree?)

BONUS: MILS

The declination above also has measurements like "70 MILS". Mils are "milliradians" or thousandths of a radian. Recall how we could have calculated the arctan in radians? If we had, we could then apply declination using mils and on a military compass that includes both degrees and mils, we could calculate our azimuth in mils.

There are 360 degrees or 6400 mils in a full circle, so the ratio is 1 degree is about 17.78 mils. For land navigation with a compass, mils are not going to be more accurate than degrees. It requires a lot of precision for mils to be more useful, such as modern artillery. It would not be bad to be familiar with mils, but you aren't going to be less lost by using mils over degrees.

CHAPTER 7

What's Next?

There is more. Isn't there always?

However, this seems like enough for the scope of this book. Here are some topics not covered here, but would be good for the avid reader to look at next.

NAVIGATION ACROSS BOUNDARIES

In the chapter on Navigation, we noted that the calculations for distance and azimuth work when the starting and ending points are in the same grid square. Then, we only need the numerical locations and the math is pretty easy. Since the grid squares aren't really squares, as we learned in the section on Grid Zone Boundaries, calculating distance and azimuth across the boundaries will require figuring out where the line crosses those boundaries first. The azimuth only needs to be calculated once, using a location within the starting grid square, but distance will need one calculation for each grid square crossed.

We would calculate the distance to the boundary, then add the distance from there to the next boundary in the adjacent grid square, and continue that until we have all grid square distances handled.

POLAR ZONES

In the section on Grid Zones, we discussed that the latitude bands use letters C through X, skipping I and O. A and B are more like semicircles attached at a line that passes through the south pole. Y and Z are the same, but for the north. The system for finding locations in the extreme north and south are similar, but are a little quirky because of the shapes. If you have cause to go to either pole, that might be worth learning more about. Also, find a good coat.

USING PAPER MAPS

The map images in this book came from screenshots of map.army. While everything we discussed applies to paper maps the same, there are tools, like a map protractor, that are a little difficult to explain in text form. Anyway, MapTools, who make the protractors I use and like best, have some good tutorials online for using a map protractor.

One insight that might not be clear at first is that the marks for eastings in the slots for measuring go from right to left. This seems wrong based on intuition, but once you use one, it will suddenly make sense. It allows you to pinpoint a location using the numerical coordinates by finding your easting and northing at the same time rather than the less accurate process if you measured them separately.

FINAL NOTE

I probably missed stuff. If so, or even if you have questions or suggestions or kudos, reach out. Email me at overton@heatherstone.com and put "MGRS Explained" in the subject line. That will help me notice your message somewhere in the endless spam that clogs my inbox daily (so feel free to try me more than once if you don't get a reply, it's not you, it's the spammers.)

Aaron "Trey" Overton keeps doing outdoor stuff. He spent all his youth in Boy Scouts, and today is both a Venturing Crew Advisor and Assistant Scoutmaster. He is also an active officer in the Tennessee State Guard.

Professionally, Trey is a software architect by trade and CTO by title. Along the way, he also taught for a bit in a high school. Teaching, training, and mentoring are part of his work and play. This is his first book on an educational topic, though you can also find his prior publications of miniature wargames, like Shock Force and GWAR: Rumble in Antarctica if you look around a bit.

It would be little effort to get him to write more books. Want to learn a thing? Ask!

www.ingramcontent.com/pod-product-compliance
Lightning Source LLC
Chambersburg PA
CBHW050817160726
48004CB00002B/879